Techniques of Safe & Vault MANIPULATION

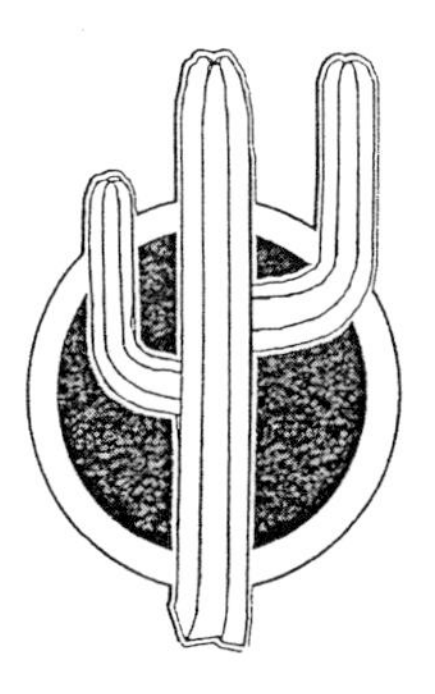

Desert Publications
El Dorado, AR 71730 U. S. A

Techniques of Safe and Vault Manipulation

505 East 5th Street
El Dorado, AR 71730 U.S.A.
1-870-862-3811
info@deltapress.com

Printed in U.S.A
ISBN: 978-0-87947-105-7
10 9 8 7 6 5 4

Desert Publication is a division of
The DELTA GROUP, Ltd.

Direct all inquiries & orders to the above address.

Table Of Contents

Foreward

Manipulation is the process of opening a combination lock, using only the eyes, ears and hands. Although an electronic listening device or stethoscope could be employed, this manual will teach you how to manipulate, void of all devices.

Movies and TV have created many false ideas about safe manipulation. The locksmith, detective, or spy usually sandpapers the finger tips, carefully listens as the "tumblers fall," and promptly opens the safe after dialing only one set of numbers.

Manipulation is strictly a mechanical procedure. The technique outlined in this manual utilizes the natural characteristics of the lock mechanism. These characteristics are the result of two things: the design of the mechanism and the manufacturing tolerances.

To practice these techniques, it will be necessary for you to obtain a lock the same as, or very similar to the one used for study and mount it on a small stand. With this set-up you simply follow the assignments. Just a few hours practice will allow you to gain the self-confidence of being able to manipulate.

Every locksmith should know how to manipulate or at least be familiar with the principles involved. This knowledge will allow you to handle all safe servicing and combination changing in a more professional manner. It is to that end that this manual was written.

It is dedicated to the countless locksmiths, locksmithing students and law enforcement agents who have never before had the opportunity to learn the techniques of manipulation.

Practice Lock

The practice lock used in this manual is a Sargent & Greenleaf R6735 x D3 x DR5. This is a 3 wheel, lever fence, key changing, combination lock. It was chosen for several reasons:

1. It is a commonly used lock and will be typical of a large portion of the safes you will be called upon to open.
2. It is fairly inexpensive and available from most large locksmith supply houses.
3. It is easy to mount and the combination can be quickly changed.
4. This lock lends itself to the ease of studying the principles involved in manipulation.

Fig. 1

PRACTICE LOCK

Sargent & Greenleaf — R6735 x D3 x DR5

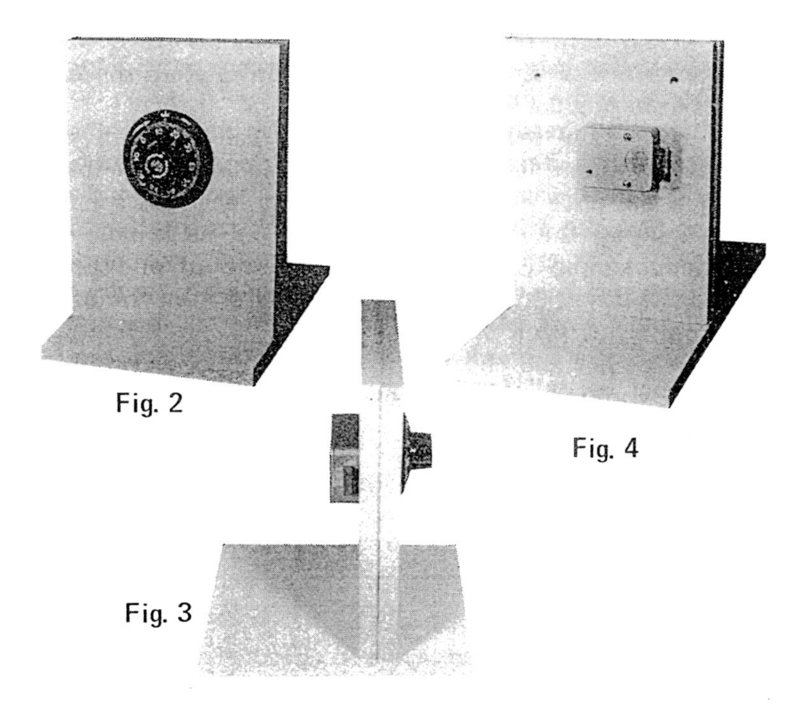

Fig. 2

Fig. 3

Fig. 4

Installation Instructions

To aid in your manipulation practice, your lock should be mounted on a small stand. The exact material and size are not important as long as the stand holds the lock in a rigid, accessible manner.

The lock shown in Figures 2, 3 and 4 was mounted on a stand made from 3 pieces of 3/4 inch particle board as used for shelving. Because of its weight, particle board makes a very solid mounting.

To mount your lock; remove the back cover, pull the spline key out, unscrew the drive cam from the spindle, pull the dial & spindle assemble out and set aside. Next remove the tube nuts and dial ring. Discard one of the tube nuts.

Your mounting stand will need a 3/4 inch diameter hole drilled through it to accept the tube. Temporarily, mount your lock as shown in Figure 6. Mark the tube, remove, and cut off 1/16 inch beyond the tube nut. File the cut end of the tube smooth and remount lock to stand.

You may now fasten the lock case to the stand with wood screws. Use care to prevent screwdriver from slipping off and damaging the lock mechanism.

You are now ready to mount the dial ring with the remaining tube nut. Reinstall the dial & spindle into the tube, screw on the drive cam, and measure the length of spindle protruding past the drive cam. Remove and cut spindle off this amount. File cut end of spindle smooth and dress threads. Re-install the dial & spindle and screw on the drive cam until tight. Back off until spline and keyway line up. Install spline key and check that the dial turns freely. Lubricate with "Lubriplate" if necessary.

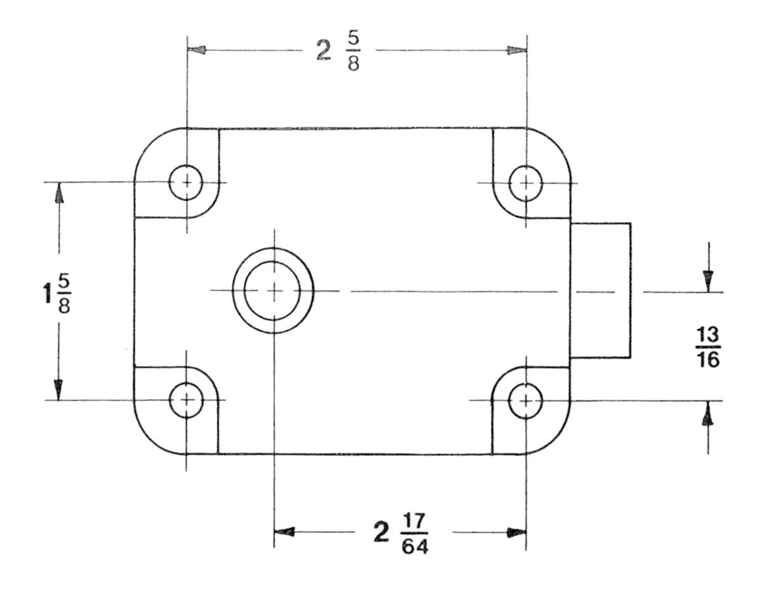
2 5/8
1 5/8
13/16
2 17/64
Fig. 5

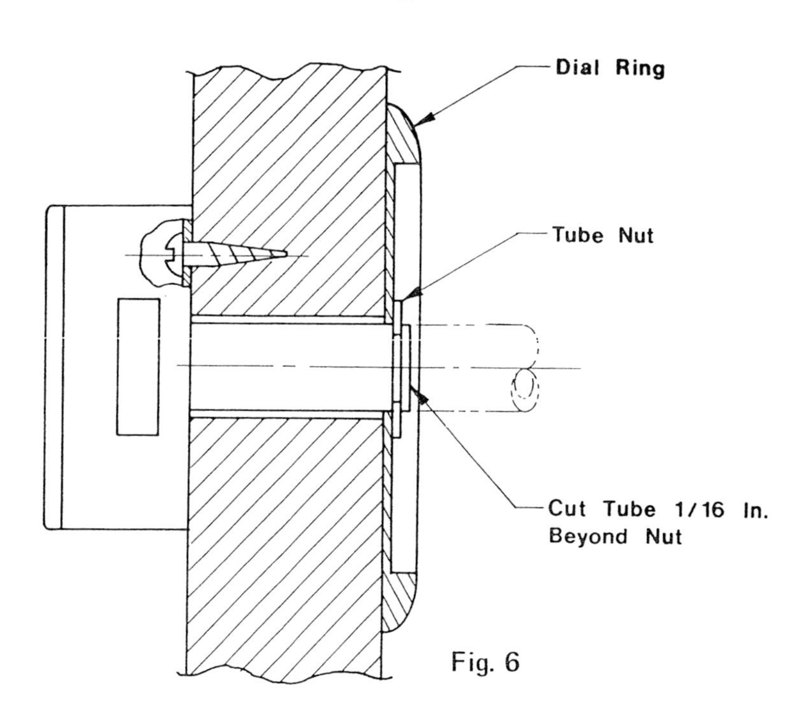
Dial Ring
Tube Nut
Cut Tube 1/16 In.
Beyond Nut
Fig. 6

How To Change Combination

SARGENT & GREENLEAF, INC.

ROCHESTER, NEW YORK

FINE LOCKS, HARDWARE AND SECURITY MECHANISMS

SINCE 1857

OPERATING AND CHANGING INSTRUCTIONS

for RIGHT HAND - VERTICAL UP - VERTICAL DOWN locks

LOCK MODELS R6720 - R6722 - R6725 - R6730 - R6732 - R6735
6220 - 6225 - 6230 - 6235

Before operating the lock or changing the combination,
READ THESE INSTRUCTIONS THOROUGHLY.

CHANGING INDEX   OPENING INDEX

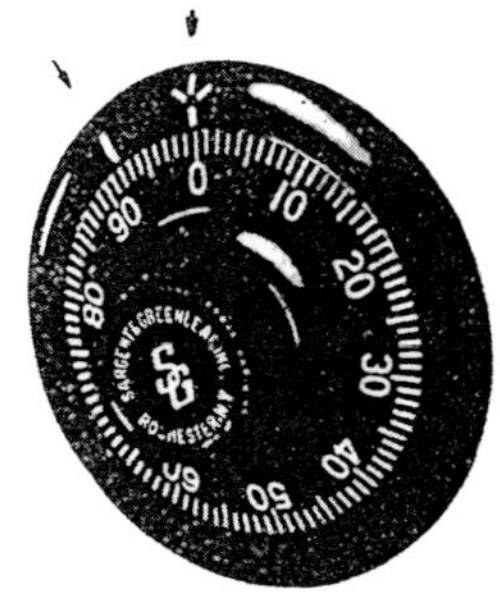

Fig. 1

At the top of the dial ring an index, , is provided for normal dialing and opening. At the side of the opening index, a changing index, I, is provided for use only when setting a new combination.

This is a precision lock, therefore extreme care must be used to align the combination numbers with the index.

Turn the dial slowly and steadily. If, after turning the correct number of revolutions, any number is turned beyond the index, , the entire series of combination numbers must be re-dialed. DO NOT TURN BACK TO REGAIN A PROPER ALIGNMENT WITH THE NUMBERS. Each time a selected number is aligned with the index a revolution is counted.

TO UNLOCK ON FACTORY SETTING

All locks in this series are set on 50 after final inspection at our factory.

To unlock when set on 50, turn dial to the LEFT four full revolutions and stop when 50 is aligned with the index, , then turn slowly to RIGHT until it stops.

TO UNLOCK ON A 3 NUMBER COMBINATION

FOR EXAMPLE 28 - 76 - 34

To unlock a lock set on a 3 number combination: For example 28-76-34.

1. Turn dial to the LEFT, stopping when "28" is aligned with the index , the FOURTH time.
2. Turn dial to the RIGHT, stopping when "76" is aligned with the index , the THIRD time.
3. Turn dial to the LEFT, stopping when "34" is aligned with the index , the SECOND time.
4. Turn dial slowly to the RIGHT until it stops.

TO LOCK

Turn dial to the LEFT at least four full revolutions.

CHANGING TO A NEW COMBINATION

Make up a new combination, selecting 3 sets of numbers of your own choosing.

Do not use numbers between 0 and 20 for your last number. (e.g. 46-82-13).

For maximum security, do not use numbers ending in 0 or 5 and do not use numbers in a rising or falling sequence. e.g. 35-50-75 is not as good a combination as 54-38-72.

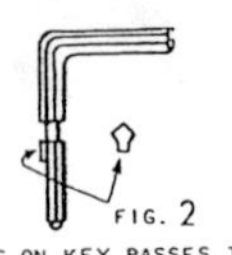

FIG. 2

WING ON KEY PASSES THRU SLOT IN LOCK PLATE.

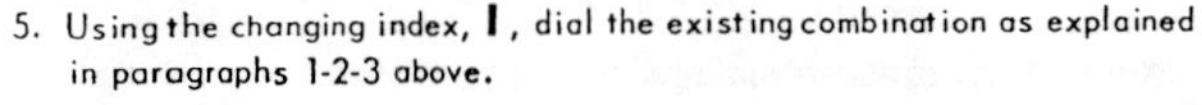

5. Using the changing index, **I**, dial the existing combination as explained in paragraphs 1-2-3 above.

   The lock leaves the factory with all 3 numbers of the combination set on 50. When setting combination for the first time: Turn dial LEFT stopping when 50 is aligned with the changing index **I**, the FOURTH time.

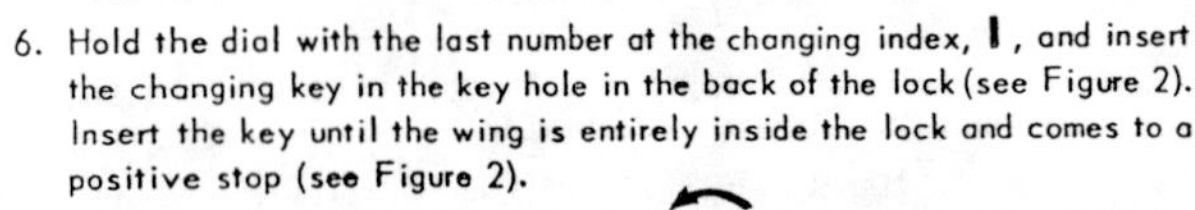

6. Hold the dial with the last number at the changing index, **I**, and insert the changing key in the key hole in the back of the lock (see Figure 2). Insert the key until the wing is entirely inside the lock and comes to a positive stop (see Figure 2).

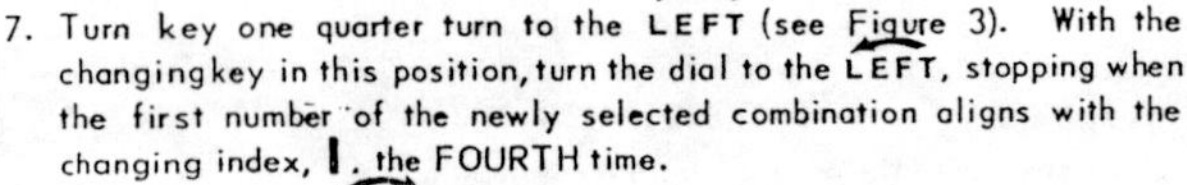

7. Turn key one quarter turn to the **LEFT** (see Figure 3). With the changing key in this position, turn the dial to the **LEFT**, stopping when the first number of the newly selected combination aligns with the changing index, **I**, the FOURTH time.

FIG. 3

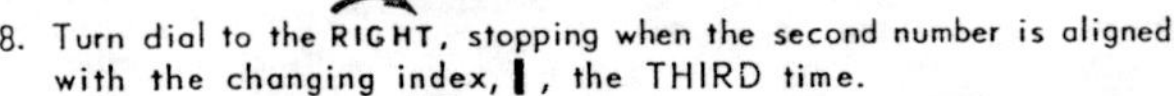

8. Turn dial to the **RIGHT**, stopping when the second number is aligned with the changing index, **I**, the THIRD time.

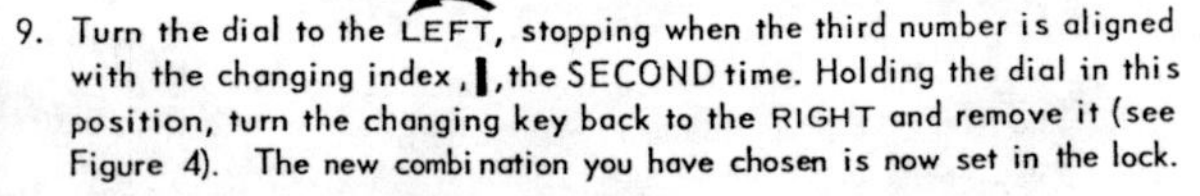

9. Turn the dial to the LEFT, stopping when the third number is aligned with the changing index, **I**, the SECOND time. Holding the dial in this position, turn the changing key back to the RIGHT and remove it (see Figure 4). The new combination you have chosen is now set in the lock.

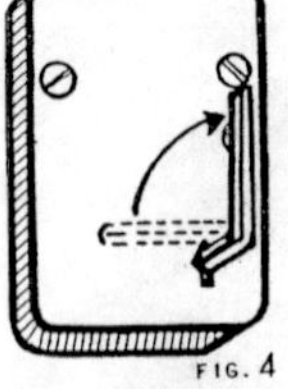

FIG. 4

CAUTION

BEFORE CLOSING AND WORKING THE UNIT,

Try the new combination several times, using the index.

WARNING: NEVER INSERT THE CHANGING KEY IN THE LOCK WHEN THE COVER IS REMOVED. ALWAYS BE CERTAIN THAT THE WING OF THE CHANGING KEY IS ENTIRELY WITHIN THE LOCK (FIGURE 2) BEFORE TURNING KEY.

IF AN ERROR HAS BEEN MADE IN SETTING A NEW COMBINATION, we suggest that an accredited locksmith be called. If this is not possible, do the following:

A. Remove the two screws on the back of the lock which hold the cover (changing key may be used as a screw driver). Remove cover.

B. Using a straightened paper clip or similar instrument, insert it in the square keyways or in the square slots in the wheels. Rotate each wheel until all the slots are in perfect alignment and the square keyways are directly over a small hole in the bottom of the case. Replace cover and insert change key. Replace screws.

C. Follow directions for setting the new combination as indicated in paragraphs 7 through 9 above.

Parts Breakdown

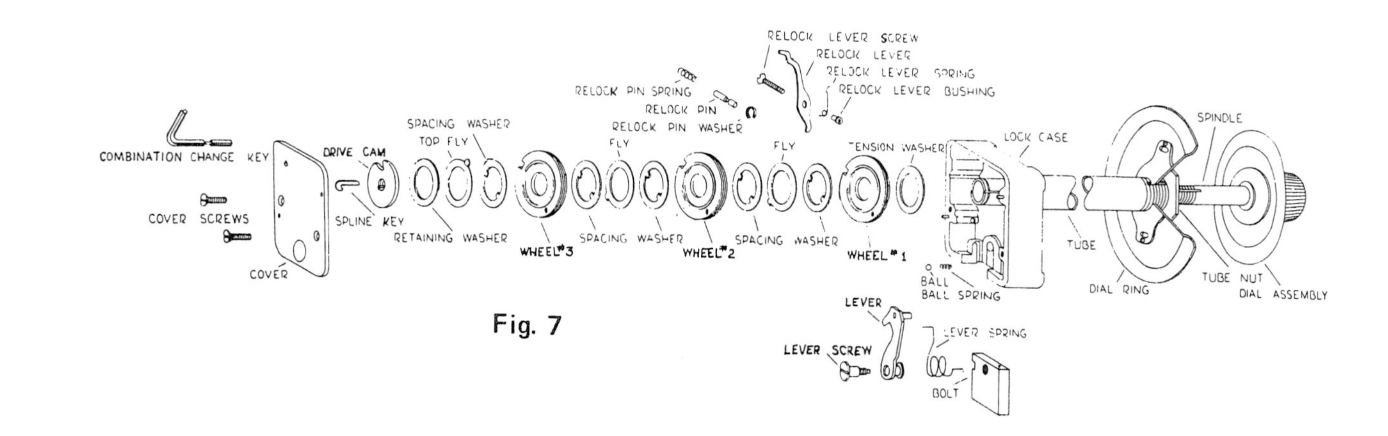

Fig. 7

How The Mechanism Works

Fig. 8

In the above photograph (Figure 8) the combination has been dialed and the dial returned to the drop-in position. With the gates of the 3 wheels lined up like this, the fence is free to drop in, allowing the nose of the lever to engage the drive cam.

Fig. 9

With the mechanism in the configuration of Figure 8, a clockwise turn of the dial will retract the bolt. Note: With the cover removed it is necessary to manually disengage the relock lever before retracting the bolt. (See Fig. 9.)

How Manipulation Is Possible

Our first exercise will show you why it is possible to manipulate this lock. Please refer to Figure 10 on page 17.

With the cover plate still removed, look at the lock mechanism. Move the dial and wheels so all the gates (b) are away from the fence (c). In this position the lever nose (d) contacts the drive cam (a), holding the fence from contacting the wheels.

Now rotate the dial & drive cam, without moving the wheels, until the dorp-in position is reached. The lever is now lower and the fence contacts the wheels.

Any variation in wheel diameters will vary the amount that the lever nose protrudes into the drop-in opening of the drive cam. Any differences in the lever travel can be read on the dial by turning it to a contact point and taking a reading.

We will now study another characteristic that makes manipulation possible. The lever (e) pivots on a shoulder screw (f). To permit it to freely pivot there must be clearance between the hole in the lever and the diameter of the screw. This clearance also allows the lever to have some sideways wobble.

We will now demonstrate just how this helps in our manipulating technique.

Again, set up the lock mechanism with all gates away from the fence, with the lever nose (d) in the drop-in opening of the drive cam, and the fence resting on the wheels. With this set-up take a reading on both contact points as shown in Figures 11, 12, 13 and 14. Write these reading down.

Now, with your finger or a small screwdriver, rotate the 3rd wheel only until its gate aligns with the fence. You should now be able to visually detect the lever twisting slightly, allowing the nose to drop a little further into the drive cam opening.

Take another set of readings on the contact points and compare with your first set. One will be lower and the other higher. Usually the contact point as shown in Figures 13 and 14 will prove to be the best to use for manipulation.

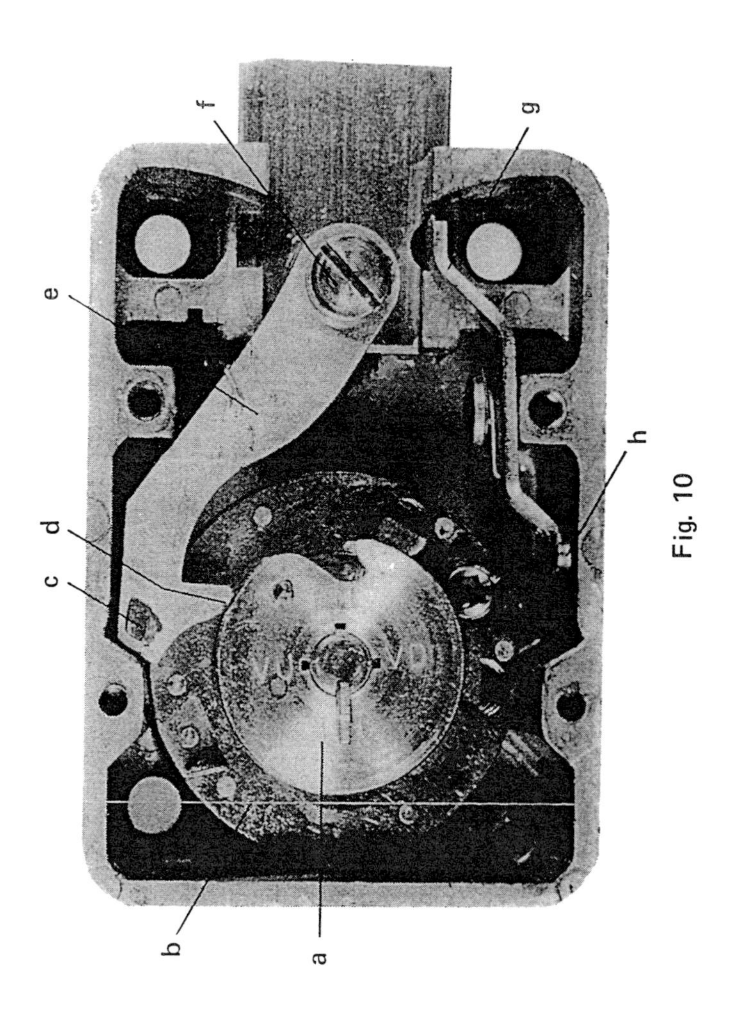

Fig. 10

How A Contact Point Reading Is Made

The photograph in Figure 11 shows one of the two contact points on this lock. This particular set-up gave a dial reading of 7 3/4. (See Figure 12) The fence is resting against the wheels. If the fence were to drop deeper into the wheels it would result in a higher reading on the dial, perhaps 7 7/8 or 8.

The dial shown in Figure 12 indicates 7 3/4. The technique of manipulation is dependant upon you training your eye to detect differences as small as 1/8 of a dial division.

The photograph (Figure 13) shows the other contact point for this lock. The lever is resting on the wheels as it was in Figure 11. This time the dial was turned clockwise to reach the contact point. The dial now reads 15. If the lever was to drop deeper the dial reading would be lower, perhaps 17 7/8.

Usually this contact point will give a better reading than the other point.

The dial shown in Figure 14 indicates 15. First train your eye to read 1/2 of a division, then 1/4, and next 1/8. If you really try, it is possible to detect 1/16 of a division difference.

Fig. 11

Fig. 12

Fig. 13

Fig. 14

Manipulation Procedure

The first portions of these exercises are done with the cover plate removed from your lock. This will better enable you to study what is happining in the mechanism as you do these exercises.

After you have studied the lock with the cover plate removed and feel competent at finding the contact points, you should proceed to learn how to determine the number of wheels the lock contains.

The drawing in Figure 15 illustrates how the dial is connected directly to the drive cam. After one complete turn, it contacts the 3rd wheel and starts it turning. Another complete turn and the 3rd wheel contacts the 2nd wheel and starts it turning. On the third and last turn the 2nd wheel contacts the 1st and now we have all 3 wheels, drive cam and dial turning as a unit.

Now, looking at just the dial, turn it at least 4 times to the right stopping at -0-. Now turn left to 90, then procede quickly past -0-. As you pass -0- you should both hear and feel the drive cam as it contacts the 3rd wheel. Continue turning to the left, again to 90, then quickly past -0-. You should have felt the 3rd wheel contact the 2nd wheel. Repeat this procedure again and you should feel the 2nd wheel contact the 1st wheel.

In this lock, the preceding procedure, repeated the fourth time, will not produce any feel or sound as the dial passes -0-. This is the indication of a 3 wheel mechanism.

The whole idea is to repeat the turning past -0- until no more wheels are indicated. It is of prime importance in manipulating to know how many wheels the lock contains.

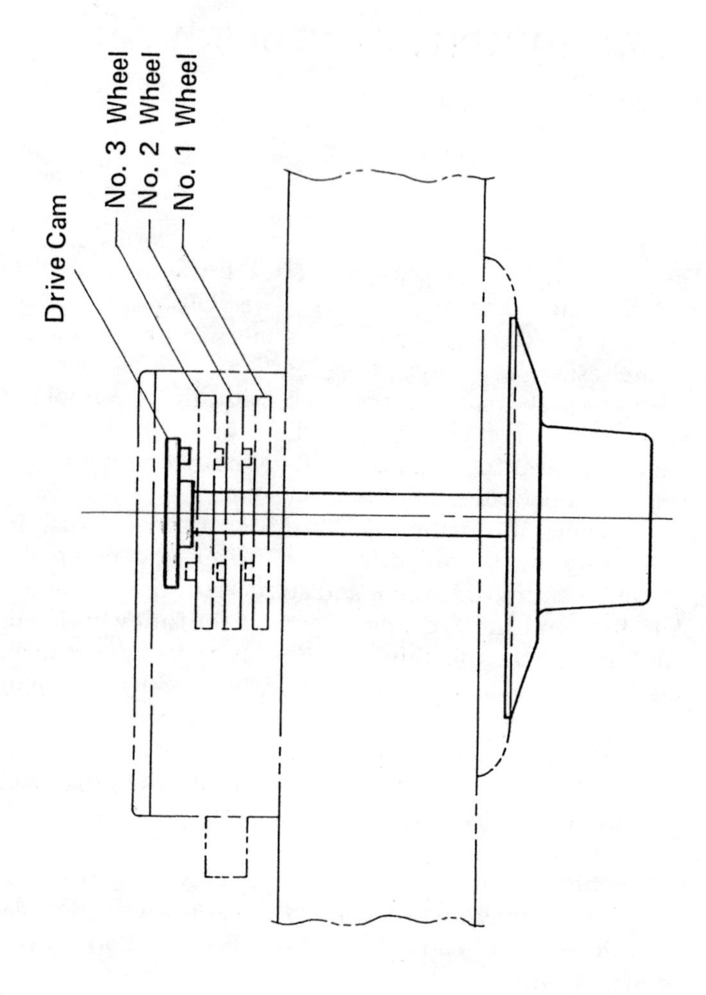

Fig. 15

Making A Graph For The 1st Number

At this stage you should feel competent to proceed to the actual manipulation procedure. If so, install the cover plate on the lock case and have someone else reset the combination so you do not know the numbers. Follow the changing instructions that came with your lock or refer to the changing instructions given earlier in this book.

Obtain some graph paper ruled similar to the sample graph shown in this book. We will be recording a contact reading every 2 1/2 numbers so your graph should accommodate 40 such readings.

In our sample graph we used the contact point as described for Figure 13. On our dial this gave a reading near 65. The drive cam can be keyed to the spindle in 4 different positions and your dial might read one of 4 different numbers, all 90° apart on the dial

Begin by turning the dial left at least 4 times until all the wheels are turning. Stop at -0-. Turn right to the contact point and record the dial reading.

Turn back left, past -0-, stopping at 2 1/2.

Turn right to contact point and record reading.

Turn back left past 2 1/2, stopping at 5.

Turn right to contact point and record the reading.

Repeat until you have moved the wheels one revolution to the left, recording the contact point every 2 1/2 numbers.

When going past the contact point you will have only a short distance to turn right for a contact point reading. When we reached 67 1/2 it was only necessary to turn right approximately 2 1/2 numbers.

Study the sample graph for finding the first number. Note that it indicates one low area in the wheel assembly between 10 and 20, a sharp decline at 55, and a moderate decline at 42 1/2. This proves to be very revealing in light of the fact that the lock combination was set to 43 — 82 — 56.

This graph indicates that very near 55 there could be a gate position for one of the 3 wheels.

We now make a "magnifying graph" to more closely examine this particular area. (See the lower line on the sample graph) Here, we want to examine the entire area between 52 1/2 and 57 1/2. Using the procedures outlined earlier, we repeat on numbers 52 through 58 to get a more precise reading. This graph now shows 56 to be a gate position for one of the wheels. At this stage we don't know which one.

To find out which wheel is indicating we use the following process of elimination:

Turn left at least 4 times, stopping at 56.

Turn right one turn, picking up the 3rd wheel at 56 and carry it past 56 about 10 numbers to approximately 46.

Turn left to the contact point and take a reading.

In this case our reading was 65 1/8 as opposed to 64 3/4, which means the 3rd wheel was the one indicating.

If the contact point still indicated low (64 3/4) we could have assumed that either the 1st or 2nd was the one indicating. The procedure then is the same as we just did, only, after picking up the 3rd wheel, turn one more revolution to the right and pick up the 2nd wheel, also. Carry both the 3rd and 2nd wheel beyond 56 to 46. Turn back left to the contact point and take a reading. This time a high reading would mean that the 2nd wheel was the one indicating. If the reading was still low, then the 1st wheel is the one indicating.

Returning to our sample first number graph, we have determined the last combination number to most likely be 56. Now, we must make another graph to plot the readings on the 1st and 2nd wheels.

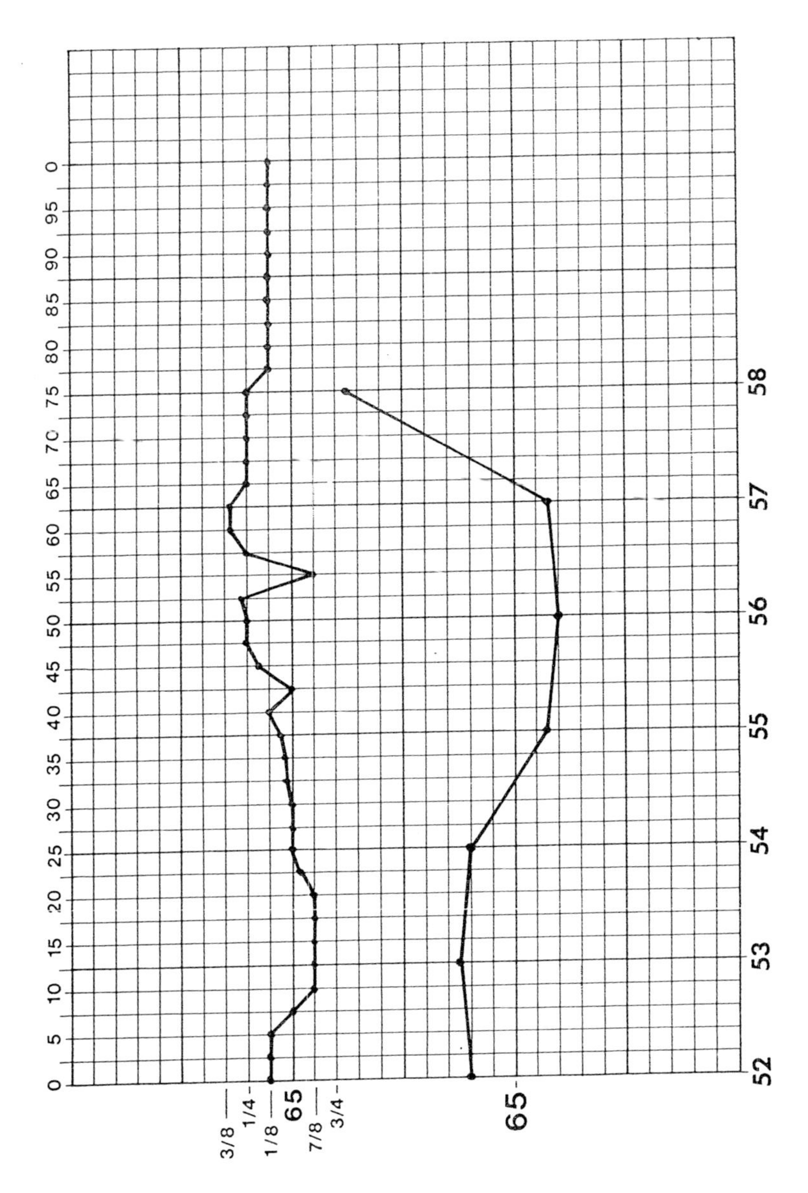

GRAPH FOR FINDING FIRST NUMBER

Making A Graph For The 2nd Number

We, now, want to read the 1st and 2nd wheels every 2 1/2 numbers with the 3rd wheel always set to the combination number we just determined (in this case 56).

Turn right at least 4 times and stop at -0-.

Turn left 1 turn, picking up the 3rd wheel at -0- and stop at 56.

Turn right to the contact point and record the reading on your second number graph.

Repeat the above procedure until you have completed a graph of every 2 1/2 numbers around the dial.

Our sample second number graph has a pronounced dip at 82 1/2 so we now do another "magnifying graph" of this particular area just as we did on our first graph. The graph now shows 82 to be the best indicating number.

We still don't know if the 1st or 2nd wheel is the one indicating 82. We must determine this in a similar manner as we did for our first number.

In the following process of elimination we will set the 1st wheel on 82, move the 2nd wheel off of 82 and set the 3rd wheel to 56. If our contact point reading is still low then the 1st wheel is indicating. If the reading is high then it is the 2nd wheel indicating. Proceed as follows:

Turn the dial at least 4 times to the left, stopping at 82.

Turn right one turn, pick up the 3rd wheel at 82 and continue right one more turn where we now pick up the 2nd wheel at 82. The 2nd wheel is carried past 82, approximately 10 numbers, to about 72.

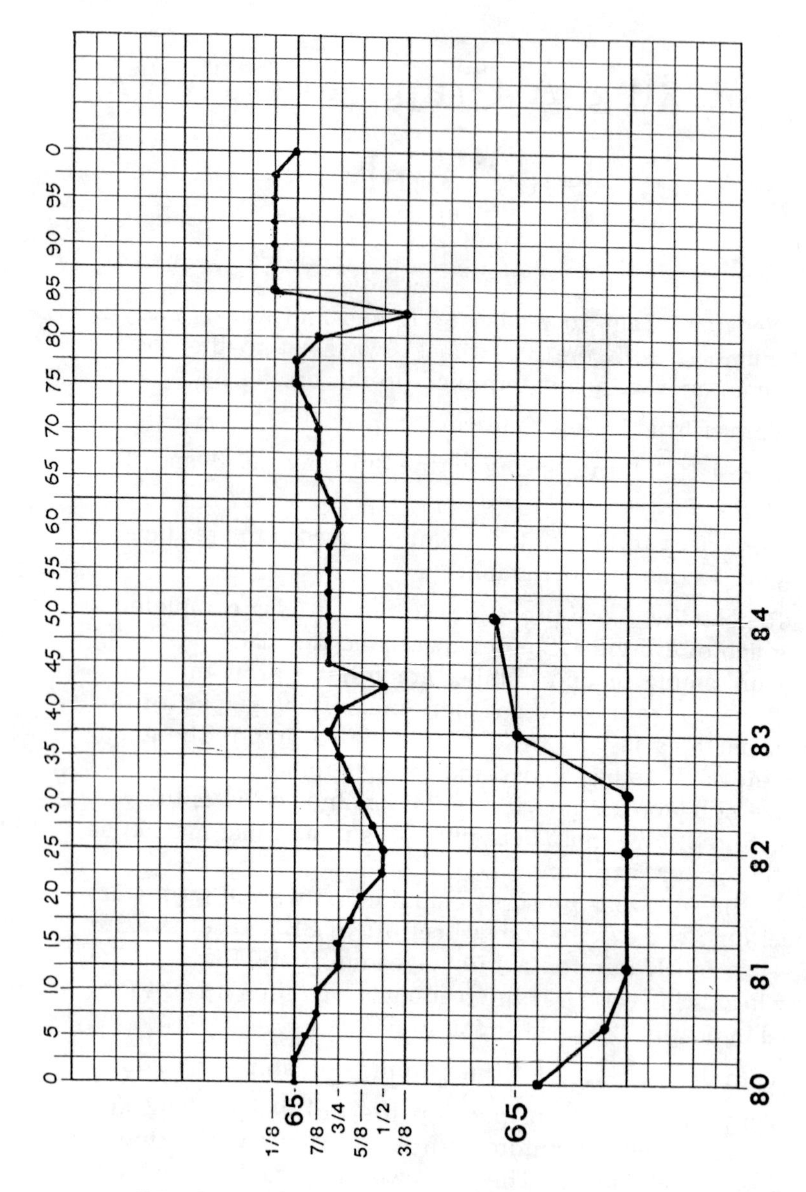

GRAPH FOR FINDING SECOND NUMBER

Turn left one turn where we pick up the 3rd wheel at 72 and continue on to 56.

Turn right to the contact point and take a dial reading.

In the case of our sample graph it was the 2nd wheel that proved to be the one indicating 82. We now know that the second combination number is 82 and the last one is 56.

The first number is now found by simply trying every 2 1/2 numbers as a possible 1st number and following through with 82 and 56 for the second and third numbers. Proceed as follows:

Turn the dial at least 4 times to the left, stopping at -0-.

Turn right past 82 two turns and stop on 82 the third time.

Turn left past 56 and stop on 56 the next time.

Turn back right to the drop-in position. Quickly twist the dial between the 2 contact points. In cases where the combination is close but not quite on, this will tend to work the fence into the gates.

Repeat this procedure, only this time set the first number on 2 1/2. Continue dialing this combination, advancing the first number 2 1/2 numbers each time until the lock opens.

The 3rd wheel will usually indicate first. If the 2nd wheel should indicate first, place it on the number indicated and make a graph on the 3rd wheel.

If the 1st wheel indicates first, place it on the number indicated and make a graph on the 2nd and 3rd wheels together. With the 1st wheel indicating first, the 2nd wheel will most likely be the next to indicate because the fence will be tilting slightly lower toward the 1st wheel.

Manipulation Proof Locks

The term "manipulation proof" is sometimes a misnomer, the same as the term "pick proof." Manufacturers choose these descriptions because they obviously have more sales appeal. A better description might be "manipulation resistant."

A manipulation resistant combination lock is a lock where the manufacturer has:

1. Tightened up manufacturing tolerances.
2. Added mechanical features to prevent the reading of contact points.
3. Added mechanical features to cause false sounds or feel.

These additions do not always make the lock totally manipulation proof. However, they sometimes make the job so slow and difficult that manipulating becomes impractical.

We shall examine one truly manipulation proof lock a little closer. The one chosen is a Sargent & Greenleaf T8415MP x D37 x DR5. This lock is a "city cousin" of the lock you have been practicing with.

The only difference in the outward appearance of this lock is the small pointer knob in the center of the dial.

Study the photographs in Figures 16, 17 and 18. The main physical difference is in the drive cam. The drive cam we were working with was a simple, one piece affair. This cam has 2 moving parts plus a spring added. We will call these parts — the inner slide (a), and the outer slide (b). These slides are actuated by a shaft extending out through the center of the spindle to a small pointed knob located in the center of the dial knob.

This lock is normally opened by dialing the combination, turning the dial to -0- (the drop-in position), and turning the pointed knob to the right. This withdraws the inner slide exposing the drop-in opening in the drive cam. The fence drops into the gates, allowing the lever nose to engage the drive cam. A clockwise turn of the dial will withdraw the bolt (See Figure 17). In Figure 18 the outer slide has been manually forced down to reveal the withdrawn inner slide.

Fig. 16

Figure 16 shows the cam slides in their relaxed position. You will note that the inner slide (a) forms a curved surface, flush with the outer diameter of the cam. The lever nose (c) now has a smooth surface to ride on and the contact points can not be felt.

Study the photograph in Figure 18 and compare it with the photograph in Figure 16. You will notice that this lock has all the features for contact point reading except they are hidden by the inner cam slide (a). The outer slide has a tab that matches a radial groove cast into the cover plate (not shown). This restricts the actuation of the cam slide mechanism except then the dial is set on -0-.

This lock could be manipulated very similar to your practice lock with one main exception. To expose the drive cam and what could be contact points you must return to -0- and withdraw the inner cam slide with the pointed knob in

Fig. 17

Fig. 18

the center of the dial. You could now turn the dial counter-clockwise and "feel" a contact point if it were not for yet another feature of this lock — the radial groove in the cover plate restricts dial rotation by means of the tab on the outer slide. This restricted dial rotation, while sufficient to withdraw the bolt when the proper combination has been dialed, does not allow the nose of the lever to touch the normal contact points.

Summation

The purpose of this manual has been to teach the basics of manipulation. Therefore, we have purposely kept to one type of lock and manufacturer for simplicity and ease of teaching.

Our recommendations are for you to keep up your practice on this one lock until you feel totally competent of your manipulation ability. Have someone else change the combinations for you and manipulate again. After thorough practice with this lock you may wish to obtain locks of other manufacturers and design to study and practice on.

You will find that different manufacturers utilize different mechanism designs which in turn demand different manipulation techniques. To have covered all types would have created a lengthy, complicated and costly manual.

It is our belief that anyone who becomes proficient at manipulating a lever-fence type lock will possess the knowledge and ability to apply these techniques to other locks.

As you progress into actual safe repair and opening you might wish to obtain a copy the "Safe & Vault Manual" available from the publisher of this manual.

Manipulation is a manual skill that places a high demand on 3 of your senses — hearing, seeing and feeling. Learning requires both study and practice. Studying this manual will give you the knowledge and practicing will give you the ability.

Appendix A

Electronic Stethoscope

The heart of this electronic stethoscope is the amplifier/speaker. This is a self-contained unit, readily available, that, with a slight modification, makes a beautiful unit. Handsomely styled and slightly larger than a package of ciratettes, it will slip into your shirt pocket when being used.

The pictures in figures 21 & 22 show the addition of a 1/8 inch miniature phono jack. This allows headphones to be used. If you so desire you can use the amplifier/speaker unit "as is" by listening to its enclosed speaker.

The headphones and harmonica mike both come equipped with 1/4 inch phone plugs. These must be replaced with miniature phone plugs. Carefully unsolder the 1/4 inch plugs and resolder the new plugs as shown in figure 29.

The construction of this electronic stethoscope is divided into 3 projects: Installing the smaller phone plugs on the headphones and mike; Altering the amplifier/speaker unit to allow use of headphones; Construction of the pickup unit.

It is assumed that those building this unit will have experience making delicate electronic solder connections. If you don't then you may want to practice some before attempting the real thing. One of the most important things is to use rosin core solder. This may sound elementary to you electronic buffs, but the author has seen complete, expensive electronic kits put together with acid core solder!!! Needless to say, the complete project ended in a self-corroding mess. If you are not sure of the type of solder you have, taste it.

The amplifier/speaker modification, while somewhat delicate, is a simple chore. It requires removing the printed circuit board, by removing 3 small screws (see * on figure

Fig. 19

Fig. 20

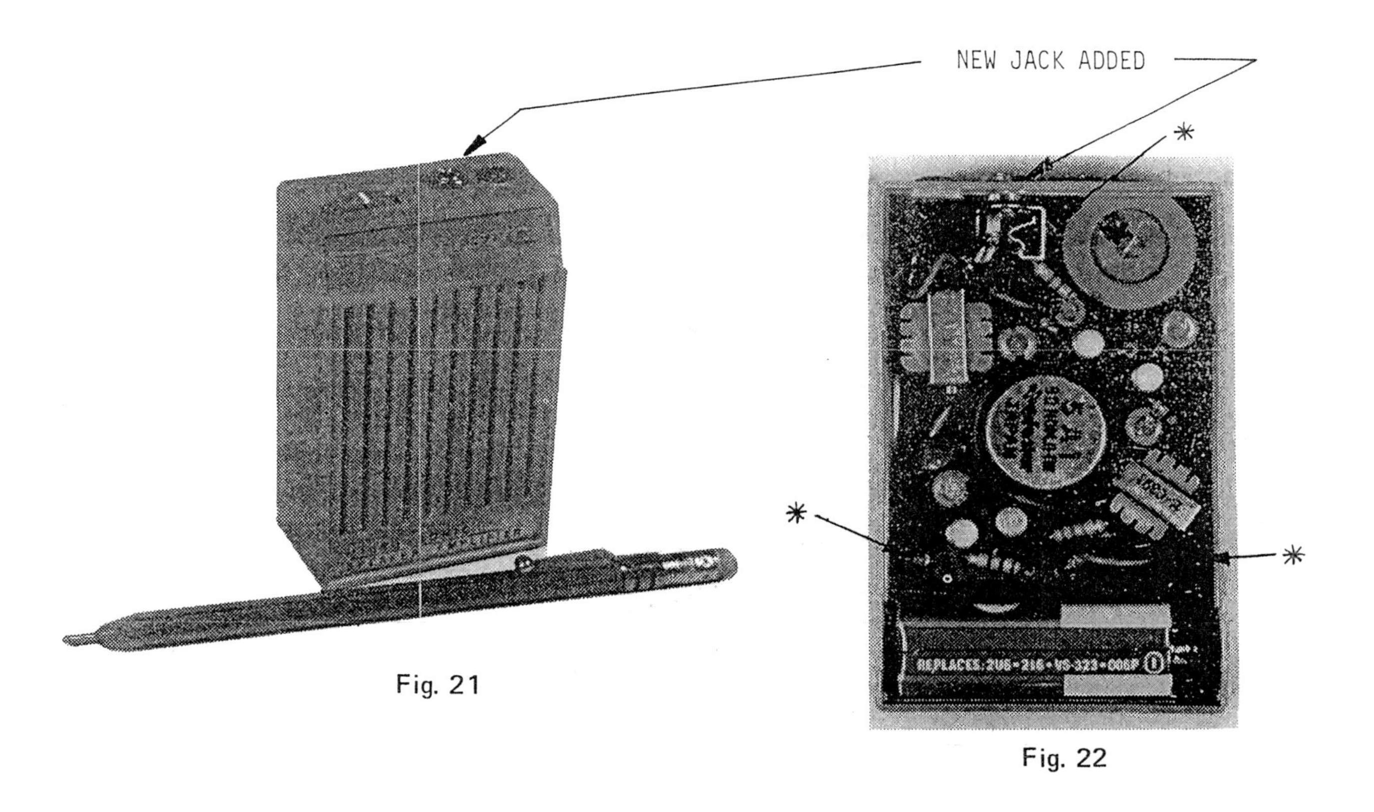

Fig. 21

Fig. 22

Fig. 23

22), removing the black wire from the speaker and soldering it to the miniature phone jack as shown in figures 25, 26 & 27. Two short pieces of hookup wire attached between the speaker and the phone jack (see figures 25, 26 & 27), completes the modification. Be sure to allow enough length in the hook-up wire to facilitate re-installing the circuit board into the case.

The pickup unit is made from two basic components: A harmonica mike and a small bakelite box. A magnetic clip (available in most five & dime stores) is altered by drilling or grinding away the rivet holding the clip to the magnet cup. The clip is discarded. The hole in the magnet cup may have to be enlarged slightly to accommodate a No. 8 - 32 screw. The head of the screw will most likely have to be machined

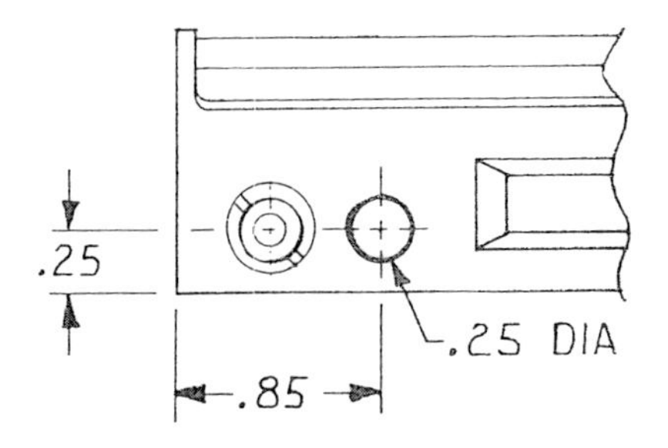

Fig. 24

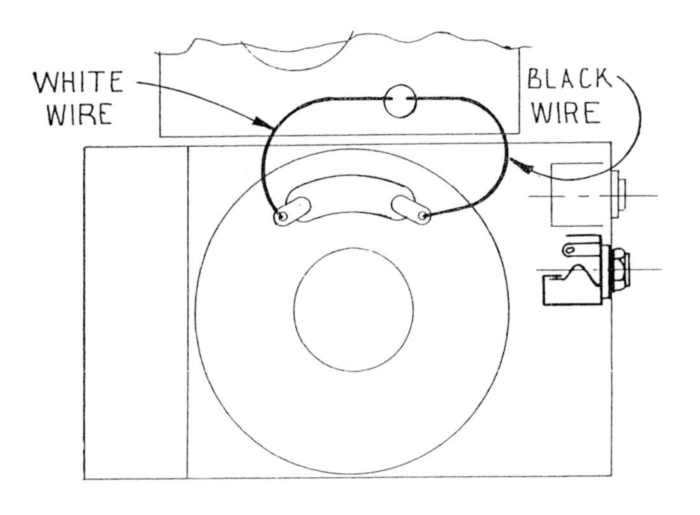

Fig. 25

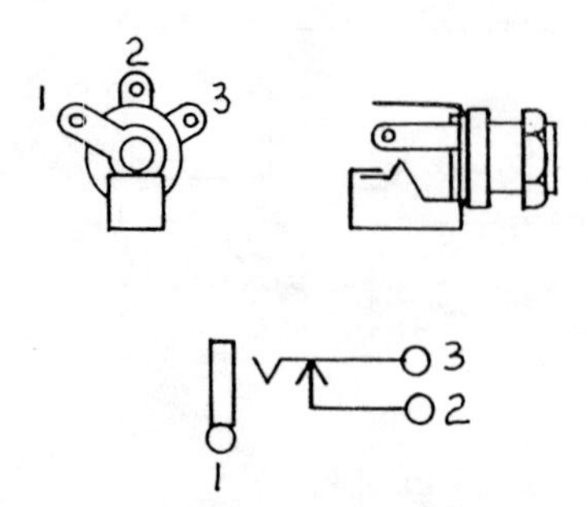

Fig. 26

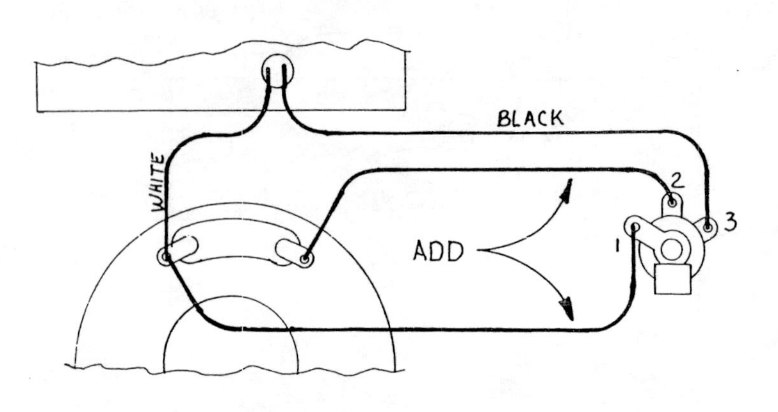

Fig. 27

Fig. 28

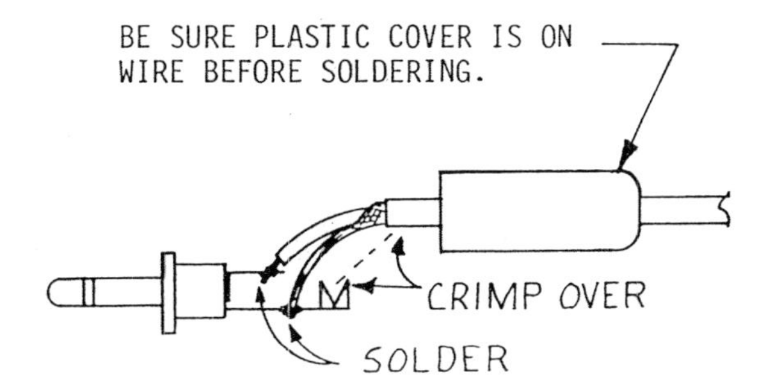
BE SURE PLASTIC COVER IS ON
WIRE BEFORE SOLDERING.
CRIMP OVER
SOLDER

Fig. 29

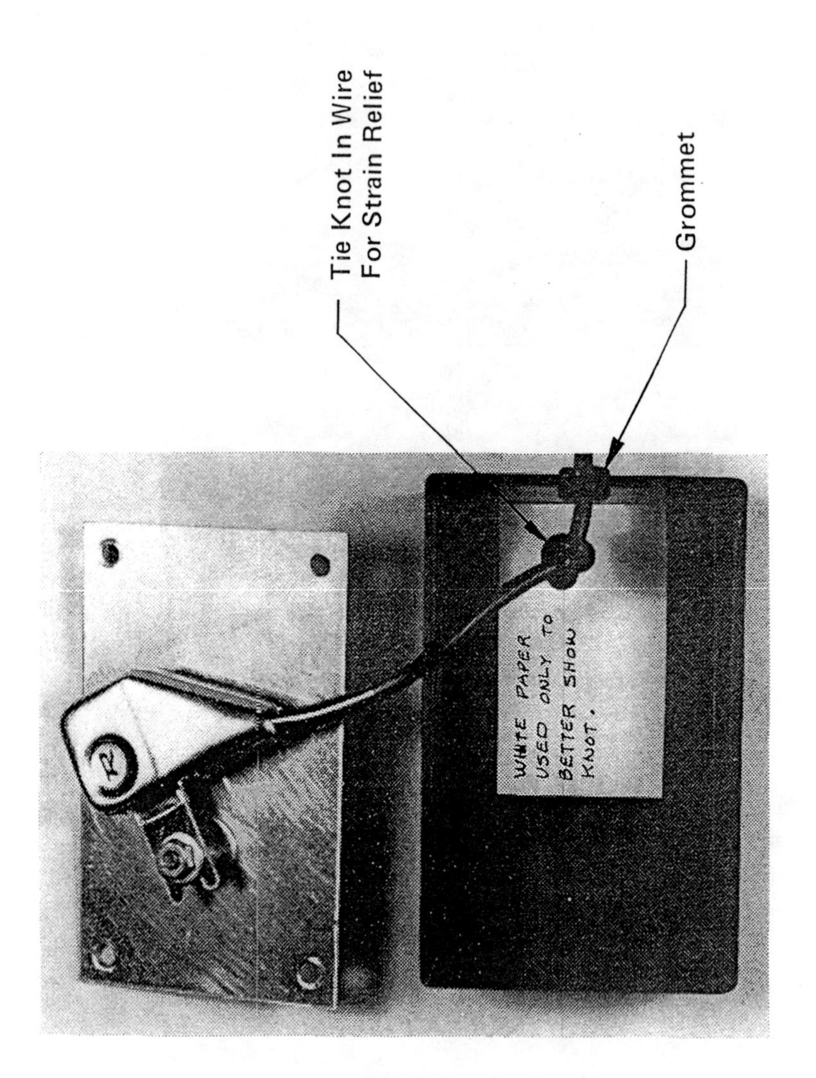
Tie Knot In Wire
For Strain Relief
Grommet
WHITE PAPER
USED ONLY TO
BETTER SHOW
KNOT.

Fig. 30

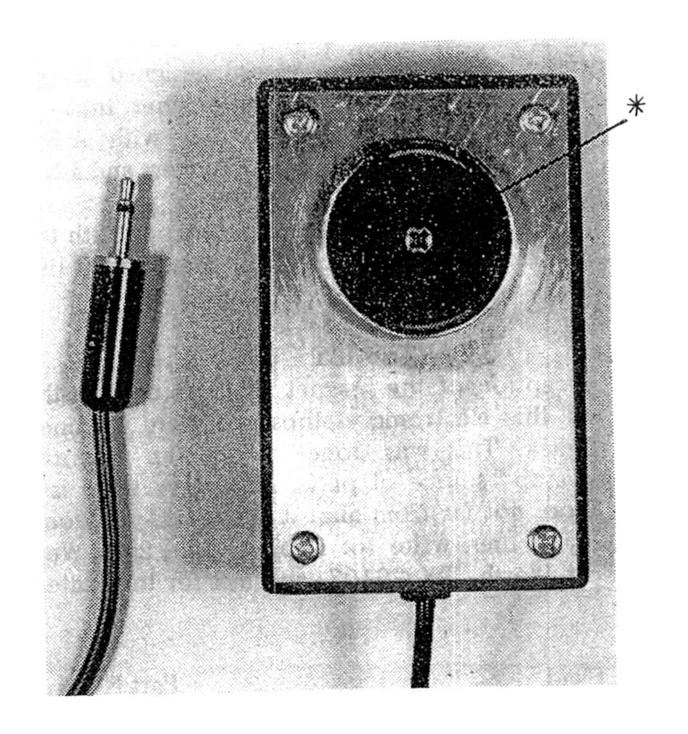

Fig. 31

down to a smaller diameter to fit into the hole in the magnet. (See * in figure 31)

Don't try to drill out the hole in the magnet. You will only end up with a broken magnet and a dull drill bit — take the word of one who knows! Machining the head of the screw is a better approach and is easily done by chucking the body of the screw in an electric drill and rotating the head against a file. This "Robinson Crusoe" method works quite well for such jobs.

The aluminum cover plate has a hole added to clear a No. 8 screw (3/16 inch diameter is OK). See figure 31 for the approximate location. The plastic box has a 1/4 inch hole drilled in one end to take a grommet (see figure 30). The exact placement of this hole is not critical.

The harmonica mike (This is a special designed crystal mike that is adapted to sense vibrations rather that air conducted sound), is fastened to the cover plate with: A No. 8 - 32 screw, 1/4 inch spacer, 2 No. 8 flat washers and a No. 8 hex nut as shown in figure 28.

The cover plate can now be attached to the box with the original screws. Your finished pickup unit should look like figure 31.

PARTS LIST

With the exception of the magnet all parts used in the construction of this electronic stethoscope were obtained from Radio Shack. This was done to give the potential builder a "one stop" source of parts. Nowadays there is a Radio Shack Store not far from almost anyone. If you don't have one near you then write to: Radio Shack, 2617 West 7th Street, Fort Worth, TX 76107, and ask for their latest catalog.

	Parts Used	Part Number
1.	Speaker/Amplifier	277 - 1008
2.	Harmonica Mike	33 - 115
3.	Headphones (8 ohm)	279 - 200
4.	Bakelite Box	270 - 230
5.	1/8" Miniature Phone Plug (Pk of 2)	274286
6.	1/8" Miniature Phone Jack Pk of 2)	274 - 253
7.	Grommet 1/8" ID (Pk of Assort. sizes)	64 - 3025
8.	Spacer (Pk of Assort. sizes)	270 - 1393
9.	Magnet (see text)	
10.	Misc. — Hardware, hook-up wire & solder.	

ELECTRONIC STETHOSCOPE ADDENDUM

Our original electronic stethoscope was designed and ɔuilt in 1975. It was designed entirely around Radio Shack :omponents in order that onyone, regardless of where he ived, could easily obtain parts, either by visiting one of their nany stores or by mail order.

Since that time Radio Shack has discontinued one vital :omponent — the harmonica mike p/n 33 - 115. This unit vas perfect for our stethoscope because it was designed to ence mechanical vibrations. With it now unavailable we had o go "back to the drawing board" and come up with a nodified pickup unit.

Our new pickup unit uses a small crystal lapel mike Radio Shack p/n 33 - 100). No drilling is done on this unit ıs the magnet, mounting plate and mike are all held together vith epoxy glue. This assembly should be clamped in place see fig. 32) and allowed to setup. A small fillet of RTV is ıow applied all around the edge of the mike, completing the ıssembly.

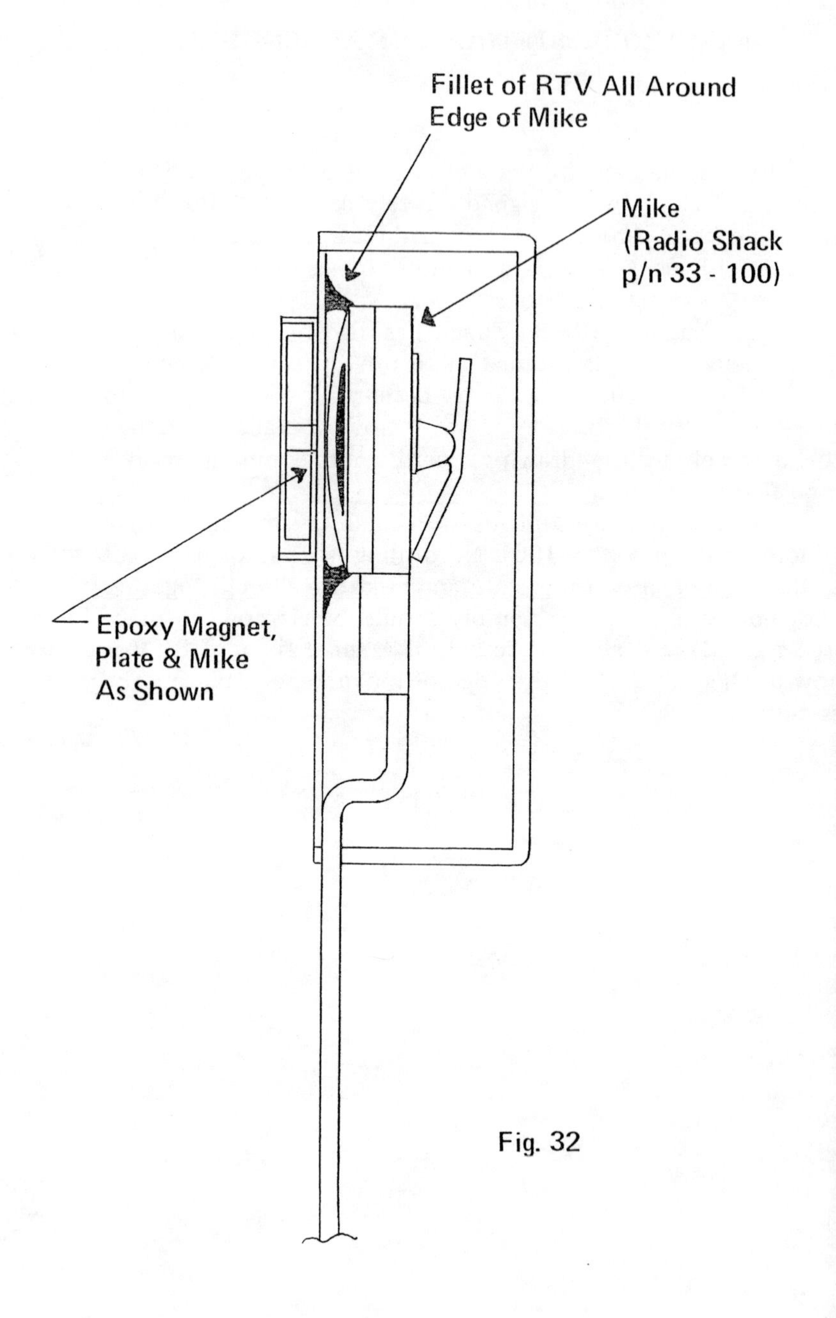
Fillet of RTV All Around
Edge of Mike
Mike
(Radio Shack
p/n 33 - 100)
Epoxy Magnet,
Plate & Mike
As Shown

Fig. 32

Other Books Available From Desert Publications

Code	Title	Price
[illegible]	Crossbows: From 35 Years With the Weapon	$15.95
[illegible]	CIA Field Expedient Preparation of Black Powder	$19.95
[illegible]	Evaluation of Improvised Shaped Charges	$19.95
[illegible]	Homeopathic First Aid	$11.95
[illegible]	Survival Childbirth	$15.95
[illegible]	How To Open A Swiss Bank Account	$9.95
[illegible]	Defending Your Retreat	$9.95
[illegible]	Methods for Long Term Storage	$9.95
[illegible]	Federal Firearms Laws	$11.95
C-024	Select Fire Uzi Modification Manual	$13.95
C-025	How to Build Silencers	$5.95
C-027	Firearms Silencers Volume 1	$13.95
[illegible]	M1 Carbine Arsenal History	$7.95
[illegible]	The Silencer Cookbook	$9.95
[illegible]	How To Build A Beer Can Mortar	$4.95
C-040	Criminal Use of False ID	$14.95
C-045	Improvised Weapons of American Underground	$14.95
C-048	Catalog of Military Suppliers	$18.95
C-050	Navy Seal Manual	$29.95
C-065	Poor Man's James Bond Vol 1	$34.95
C-113	Submachine Gun Designers Handbook	$19.95
C-114	AR-15, M16 and M16A1 5.56mm Rifles	$13.95
C-117	Ruger Carbine Cookbook	$13.95
C-120	Shootout II	$14.95
C-122	AK-47 Assault Rifle	$15.95
C-123	Combat Loads for Sniper Rifles	$18.95
C-125	Browning Hi-Power Pistols	$13.95
C-126	UZI Submachine Gun	$13.95
C-127	FullAuto Vol 1 Ar-15 Modification Manual	$11.95
C-128	Full Auto Vol 2 Uzi Mod Manual	$11.95
C-130	Full Auto Vol 4 Thompson SMG	$11.95
C-131	FullAuto Vol 5 M1 Carbine to M2	$11.95
C-135	M-14 Rifle, The	$12.95
C-136	Fighting Garand Manual	$13.95
C-137	Ranger Training & Operations	$19.95
C-138	Emergency War Surgery	$29.95
C-140	Emergency Medical Care/Disaster	$14.95
C-143	Survival Medicine	$15.95
C-150	Improvised Munitions Black Book Vol 1	$17.95
C-151	Improvised Munitions Black Book Vol 2	$17.95
C-152	Impro. Munitions Black Book Vol 3	$29.95
C-155	Fighting Back on the Job	$14.95
C-162	Colt .45 Auto Pistol Manual	$13.95
C-163	Survival Evasion & Escape	$19.95
C-165	USMC Sniping	$19.95
C-175	Beat the Box	$7.95
C-197	How to Make Disposable Silencers Vol. 1	$19.95
C-198	Expedient Hand Grenades	$21.95
C-199	U. S. Army Counterterrorism Training Manual	$19.95
C-203	Explosives and Propellants	$19.95
C-204	Improvised Munitions/Ammonium Nitrate	$19.95
C-205	Two Component High Explosives Mixtures	$19.95
C-207	CIA Improvised Sabotage Devices	$19.95
C-208	Spcial Forces Demolitions Training Handbook	$19.95
C-210	Guerilla Warfare	$17.95
C-211	FN-FAL Auto Rifles	$21.95
C-218	Professional Homemade Cherry Bomb	$15.95
C-219	Improvised Shaped Charges	$19.95
C-220	Improvised Explosives, Use In Detonation Devices	$19.95
C-227	Springfield '03 Rifle Manual	$15.95
C-229	Lock Picking Simplified	$11.95
C-230	How to Fit Keys by Impressioning	$9.95
C-231	Lockout -Techniques of Forced Entr.	$15.95
C-235	How to Open Handcuffs Without Keys	$13.95
C-244	Camouflage	$14.95
C-274	M14 and M14A1 Rifles and Rifle Marksmanship	$23.95
C-284	Thompson Submachine Guns	$19.95
C-289	M1 Carbine Owners Manual	$13.95
C-290	Clandestine Ops Man/Central America	$14.95
C-324	How to Make Disposable Silencers Vol. 2	$19.95
C-333	USMC AR-15/M-16 A2 Manual	$21.95
C-337	Sten MKII SMG Construction Manual	$24.95
C-356	Map Reading and Land Navigation	$19.95
C-365	Survival Gunsmithing	$14.95
C-372	Poor Man's James Bond Vol 2	$34.95
C-385	Self-Defense Requires No Apology	$15.95
C-414	Select Fire 10/22	$15.95
C-415	Construction Secret Hiding Places	$15.95
C-416	Brown's Alcohol Motor Fuel Cookbook	$24.95
C-417	Canteen Cup Cookery	$9.95
C-418	Improvised Batteries/Detonation Devices	$19.95
C-421	Walther P-38 Pistol Manual	$13.95
C-422	P-08 Parabellum Luger Auto Pistol	$14.95
C-423	Sten Submachine Gun, The	$11.95
C-427	How to Train a Guard Dog	$16.95
C-429	The Anarchist Handbook Vol. 1	$14.95
C-430	Beretta - 9MM M9	$15.95
C-432	Mercenary Operations Manual	$14.95
C-433	Improvised Lock Picks	$15.95
C-455	Modern Day Ninjutsu	$17.95
C-512	FullAuto Vol 8 M14A1 & Mini 14	$11.95
C-518	L.A.W. Rocket System	$14.95
C-526	With British Snipers To the Reich	$29.95
C-532	Dental Specialist	$59.95
C-553	Food, Fur and Survival	$17.95
C-561	The Squeaky Wheel	$17.95
C-562	Survival Chemist Book	$17.95
C-564	Medical Specialist	$49.95
C-571	U. S. Marine Corps Scout/Sniper Training Manual	$27.95
C-610	Aunt Bessie's Wood Stove Cookbook	$9.95
C-622	Defensive Shotgun	$14.95
C-625	MAC-10 Cookbook	$14.95
C-627	Brown's Lawsuit Cookbook	$26.95
C-628	Can You Survive	$16.95
C-633	Hand to Hand Combat	$11.95
C-634	USMC Hand to Hand Combat	$9.95
C-635	Hand to Hand Combat by D'Eliscue	$11.95
C-636	Prisons Bloody Iron	$15.95
C-639	Police Karate	$17.95
C-646	Science of Revolutionary Warfare	$16.95
C-658	Survival Shooting for Women	$7.48
C-680	Infantry Scouting, Patrol, & Sniping	$15.95
C-694	CIA Field Expedient Incendiary Manual	$19.95
C-761	How to Lose Your X-Wife Forever	$24.95
C-763	Vigilante Handbook	$19.95
C-775	Alcohol Distillers Handbook	$27.95
C-776	How to Build a Junkyard Still	$15.95
C-777	Engineer Explosives of WWI	$19.95
C-794	US Marine Corps Essential Subjects	$19.95
C-815	US Marine Bayonet Training	$11.95
C-818	The Poisoner's Handbook	$29.95
C-823	Cold Weather Survival	$14.95
C-829	The Anachist Handbook Vol. 2	$14.95
C-899	Napoleon's Maxims of War	$11.95
C-890	Company Officers HB of Ger. Army	$19.95
C-891	German Infantry Weapons Vol 1	$18.95
C-896	Op. Man. 7.62mm M24 Sniper Weapon	$13.95
C-899	US Army Bayonet Training	$11.95
C-9002	German MG-34 Machinegun Manual	$11.95
C-9012	Firearm Silencers Vol 3	$18.95
C-9013	HK Assault Rifle Systems	$27.95
C-9047	SKS Type of Carbines, The	$17.95
C-9048	Rough Riders, The	$24.95
C-9052	Private Weaponeer, The	$17.95
C-9057	Keys To Understanding Tubular Locks	$11.95
C-9060	The Anarchist Handbook Vol. 3	$14.95
C-9080	Leadership Hanbook of Small Unit Ops	$14.95
C-9081	Dirty Fighting	$13.95
C-9082	Lasers & Night Vision Devices	$29.95
C-9083	Water Survival Training	$8.95
C-9084	Assorted Nasties	$29.95
C-9085	Ruger P-85 Family of Handguns	$19.95
C-9101	Guide to Germ Warfare	$17.95
C-9102	CIA Field Expedient Method for Explosives Preparation	$19.95
C-9119	Surviving Global Slavery	$16.95
C-9120	Military Knife Fighting	$14.95
C-9127	H&R Reising Submachine Gun Manual	$19.95
C-9131	Live to Spend It	$29.95
C-9134	Military Ground Rappelling Techniques	$15.95
C-9135	Smith & Wesson Autos	$19.95
C-9137	Caching Techniques of U.S. Army Special Forces	$14.95
C-9138	USMC Battle Skills Training Manual	$49.95
C-9139	Survival Armory	$27.95
C-9156	Concealed Carry Made Easy	$17.95
C-9164	The L'il M-1, The .30 Cal. M-1 Carbine	$17.95
C-9170	Urban Combat	$27.95
C-9178	Apocalypse Tomorrow	$17.95
C-9198	MP40 Machinegun Manual	$16.95
C-9199	Clear Your Record & Own a Gun	$24.95
C-9200	Sig's Handguns Manual	$19.95
C-9210	Sniper Training	$24.95
C-9212	Poor Man's Sniper Rifle	$19.95
C-9221	The Official Makarov Pistol Manual	$14.95
C-9224	The Butane Lighter Hand Grenade	$14.95
C-9228	Unarmed Against the Knife	$14.95
C-9229	Black Book of Booby Traps	$29.95
C-9239	Militia Battle Manual	$22.95
C-9247	Black Book of Revenge	$15.95
C-9253	Bankruptcy, Credit and You	$18.95
C-9255	The Sicilian Blade	$15.95
C-9259	How to Build Practical Firearm Silencers	$18.95
C-9262	Glock's Handguns	$19.95
C-9263	Heckler and Koch's Handguns	$19.95
C-9264	The Poor Man's Ray Gun	$11.95
C-9265	The Poor Man's R. P. G.	$24.95
C-9279	Credit Improvement & Protection Handbook	$19.95
C-9281	Build Your Own AR-15	$19.95
C-9282	How to Live Safely in a Dangerous World	$19.95
C-9293	Bllack Book of Arson	$24.95
C-9322	How to Build Flash/Stun Grenades	$19.95
C-9342	Entrapment	$19.95
C-9359	The FN-FAL Rifle Et Al	$21.95
C-9367	The Mental Edge, Revised	$17.95
C-9373	Survival Bible by Duncan Long	$59.95
C-9405	The Internet Predator	$17.95
C-9406	The Criminal Defendants Bible	$49.95
C-9430	Protect Your Assets	$24.95
C-9432	How To Build Military Grade Suppressors	$21.95
C-9441	Mossberg Shotguns	$24.95
C-9478	Poor Man's Primer Manual, The	$24.95
C-9546	Agents HB of Black Bag Ops	$14.95
C-9556	Improvised Rocket Motors	$14.95
C-9612	Homestead Carpentry	$14.95

PRICES SUBJECT TO CHANGE WITHOUT NOTICE

Desert Publications

505 East Fifth St., Dept. - BK - 105
El Dorado, AR 71730
www.deltapress.com

870-862-3811
info@deltapress.com

Shipping & Handling
1 item $7.95 - 2 or more $11.95